Also by Dr. Leo Lexicon

AI for Smart Pre-Teens and Teens Ages 10-19: Using Artificial Intelligence to Learn, Think, and Create

AI for Smart Kids Ages 6-9: Discover How Artificial Intelligence is Changing the World

The AI Nerd: Quizmaster Edition Mind-Blowing AI Quizzes that Educate, Entertain and Challenge

Teen Innovators: 30 Teen Trailblazers and their Breakthrough Ideas

Innovation Handbook for Teen Entrepreneurs: Strategies, Tools and Resources to Transform your Vision into Reality

10 Life Hacks Every Teen Should Know: A Comprehensive Guide to Empowerment, Success and Fulfillment in the Teenage Years

Richard Feynman: The Adventures of a Curious Physicist

Nikola Tesla: An Electrifying Genius

John von Neumann: A Giga Brain

Elon: A Modern Renaissance Man

Einstein: The Man, The Myth, The Legend

George Washington: The First American President

Robert Falcon Scott: A Pioneer of Antarctic Exploration

Marco Polo: Intrepid Explorer who Bridged East and West

Captain Cook: The Legendary Seafarer, Navigator, and Explorer

Julius Caesar: The Rise and Fall of Rome's Greatest Leader

Cleopatra: Queen of the Nile

Frida Kahlo: Unbroken Spirit: Artist, Activist, and Icon

QUANTUM COMPUTING for Smart Pre-Teens and Teens Ages 10-19

Quantum Nerd Quizmaster Edition: Quantum Quizzes that Educate, Entertain and Challenge

CHEMISTRY NERD: 1000+ Amazing And Mind-Blowing Facts About Chemistry

PHYSICS NERD: 1000+ Amazing And Mind-Blowing Facts About Physics

BIOLOGY NERD: 1000+ Amazing And Mind-Blowing Facts About Biology

ASTRONOMY NERD: 1000+ Amazing And Mind-Blowing Facts About Astronomy

ASTRONOMY Nerd: 1000+ Amazing and Mind-Blowing Facts About our Planet, Solar System, Galaxy and Universe

© 2024 by Dr. Leo Lexicon

ASTRONOMY Nerd: 1000+ Amazing and Mind-Blowing Facts About our Planet, Solar System, Galaxy and Universe

by

Dr. Leo Lexicon

v

ASTRONOMY Nerd: 1000+ Amazing and Mind-Blowing Facts About our Planet, Solar System, Galaxy and Universe

Welcome to Astronomy Nerd!

Dive into the expansive world of "Astronomy Nerd" and join Dr. Leo Lexicon on an awe-inspiring journey through the cosmos. This essential read for any astronomy enthusiast offers a comprehensive exploration of celestial phenomena, from the familiar rocks orbiting our Sun to the enigmatic frontiers of spacetime.

In this book, you will:

- Gain a fundamental understanding of our solar system, from the Sun to the smallest meteorites.
- Explore a diverse array of celestial bodies, including planets, dwarf planets, and comets.
- Uncover the latest discoveries in our solar neighborhood that reshape our cosmic perspective.
- Delve into the varied types of stars and the historical significance of constellations.
- Learn about the recent advancements in stellar astronomy that are pushing the boundaries of our knowledge.
- Comprehend the intricate concepts of relativity and spacetime and their implications for our understanding of the universe.
- Dive into the mysterious realms of black holes and wormholes, with the latest theories bringing science fiction closer to reality.
- Investigate exoplanets and the essential conditions needed for life beyond our Earth.
- Stay updated with the newest findings in the search for extraterrestrial intelligence (SETI).
- Experience the history and excitement of rocketry and spaceflight, including a look at future exploration goals.

- Discover the roots of astronomy in ancient civilizations and its impact on art and literature.
- Look ahead to the future of space exploration with cutting-edge research and upcoming space telescopes.

"Astronomy Nerd" is a companion for your imagination and a telescope for your intellect, inviting you to join the ranks of those who keep their eyes on the stars, planets and the worlds beyond.

Dr. Leo Lexicon is an educator and author. He is the founder of Lexicon Labs, a publishing imprint that is focused on creating entertaining and educational books for active minds.

CONTENTS

Chapter 1: The Fundamentals

Overview of the Sun

We will start our astronomical journey in our own backyard – our solar system. As we already know, the Sun is the star at the center of our solar system and is by far the largest object in the solar system. With a diameter of about 1.4 million kilometers, the Sun contains 99.8% of the total mass of the entire solar system. The immense gravitational influence of the Sun holds the planets, dwarf planets, asteroids, comets and other objects in orbit around it.

But what exactly is the sun? It is classified as a G-type main sequence star, also known as a yellow dwarf star. It formed about 4.6 billion years ago from the gravitational collapse of a giant molecular cloud consisting mainly of hydrogen and helium. The temperature and pressure in the core of the Sun are extremely high, allowing thermonuclear fusion reactions to occur. This nuclear fusion converts hydrogen into helium and releases enormous amounts of energy in the form of gamma rays and other radiation.

The core of the Sun extends from the center to about 20-25% of the solar radius. It has a density of up to 150 g/cm3 and a temperature of 15 million kelvins. The proton-proton chain reaction is the predominant fusion process in the core, releasing energy and neutrinos. Surrounding the core is the radiative zone where energy is transported outward by thermal radiation. Next is the convective zone where hot loops of plasma and gas cells transport more energy outward. The visible surface of the Sun that we see is called the photosphere and has a temperature of about 5800K. Above the photosphere lie the chromosphere and the corona, which have temperatures of tens of thousands of kelvins. The immense energy output of the Sun, totaling about 3.8 x 1026 Watts, powers life on Earth along with shaping its climate and weather. Only a tiny fraction of the Sun's radiation reaches the Earth, yet that is more than 10,000 times greater than all other energy sources on Earth combined. Solar radiation takes about 8 minutes to travel the 150 million kilometers to Earth. Fluctuations in solar activity like sunspots, solar flares, and coronal mass ejections can affect space weather around Earth.

The Sun does not have a solid surface but gaseous material rotates faster at its equator than at higher latitudes. This differential rotation indicates that the Sun has a magnetic field extending outward into the solar system. The rotation period at the equator is about 25 days compared to around 35 days at the poles. The Sun's magnetic field goes through a cycle every 11 years, marked by the level of sunspot activity.

Over its 4.6-billion-year lifetime, the Sun has increased about 30% in luminosity as the helium concentration in the core builds up. As the Sun continues to fuse hydrogen into helium, models predict it will expand into a red giant in about 5 billion years and eventually blow off its outer layers as a planetary nebula. The remaining core will cool into a white dwarf. Our Sun is about halfway through the main sequence evolutionary phase for a star of its mass.

The Planets

The planets of our solar system consist of the four inner rocky planets, the four outer gas giants, and the five dwarf planets. The four inner or terrestrial planets are Mercury, Venus, Earth and Mars. They are composed mostly of metals and silicate rocks, have relatively high densities, and have solid surfaces. The four outer planets are Jupiter, Saturn, Uranus and Neptune and are composed predominantly of lighter elements like hydrogen, helium and ices. Lacking solid surfaces, they are classified as gas giants.

The first planet from the Sun, Mercury, has a heavily cratered surface with smooth flat plains. With no atmosphere to retain heat, it has extreme temperature variations from -180°C to 430°C. Venus is perpetually shrouded in thick clouds of sulfuric acid that trap heat and make it the hottest planet. Earth is the only planet known to harbor life with its nitrogen-oxygen atmosphere, oceans of liquid water, and active geology. Mars has the largest volcanoes in the solar system and evidence it once had rivers and lakes. Jupiter, the largest planet, has turbulent weather systems like its Great Red Spot, a giant ongoing storm. Saturn is distinguished by its extensive ring system made of ice and rock particles. Uranus rotates nearly on its side and has unusual retrograde motion. Neptune has extremely fast winds and Great Dark Spot storms that come and go. All four gas giants have numerous moons and rings, most notably Saturn.

The terrestrial planets all have solid surfaces with mountains, valleys, canyons, volcanoes and craters from impacts. They have metal cores and rocky mantles with varying evidence of past geological activity. Their orbits are close to circular but have eccentricities of a few percent. Orbital periods range from 88 days for Mercury to 225 days for Mars. All rotate or spin as they revolve around the Sun, with rotational periods between 24 hours and 243 days. Their axial tilts, from near 0 to 30 degrees, create seasonal climate variations.

While the inner planets formed from metal and rock that condensed near the hot young Sun, the outer planets formed in the colder outer solar system where icy materials like water, ammonia and methane were abundant. The gas giants are substantially larger than the rocky planets, with radii from 4 to 11 times Earth's radius. They have no solid surfaces but their internal structures consist of dense cores surrounded by envelopes of metallic hydrogen and helium, and finally the outer atmospheres. Their many moons are likely captured asteroids or Kuiper belt objects. Orbital periods for the giant planets range from 12 years for Jupiter to 165 years for Neptune. Rotation rates are also faster than the rocky worlds, between 10 hours to 17 hours. Their rapid spin results in flattening and oblateness at their poles.

All eight planets have poles, equators and orbital motions following Newton's laws of motion. Their orbital planes are aligned meaning they revolve around the Sun in the same direction. Most have natural satellites or moons orbiting around them. The Moon orbits Earth while the other planets except Mercury and Venus has multiple moons. Jupiter and Saturn have dozens each while Uranus has 27 known moons. There are nearly 200 moons in total known in the solar system. The planets with moons exert gravitational forces that cause tidal heating and geological activity on some of them. Several moons show evidence of subsurface oceans under frozen crusts, making them potentially habitable environments.

Dwarf Planets and Asteroids

In addition to the eight major planets, there are five recognized dwarf planets in the outer solar system. These worlds are smaller than regular planets but still orbit the Sun and are spheroid or nearly round in shape. The dwarf planets are Pluto, Ceres, Eris, Haumea and Makemake.

Pluto was discovered in 1930 and considered the ninth planet until 2006 when it was reclassified as a dwarf planet along with Ceres,

Haumea, Makemake and Eris. Located about 40 times farther from the Sun than Earth on average, frigid Pluto has a tenuous nitrogen atmosphere and a surface made of frozen nitrogen, methane and carbon monoxide ices. NASA's New Horizons spacecraft flew by Pluto in 2015 and revealed towering ice mountains, active geology and evidence of a subsurface ocean. One of Pluto's five moons, Charon, is about half its size and they orbit each other like a binary system.

The dwarf planet Ceres resides in the asteroid belt between Mars and Jupiter and is by far the largest and most massive object there at 950 km in diameter. Ceres may have a muddy ice mantle and salty ocean beneath its crust. Strong evidence of water vapor, ice and organic compounds has been found there. The Dawn spacecraft orbited Ceres from 2015-2018 and observed over 130 bright spots which are salt deposits from subsurface brine reaching the surface.

Eris is about the same size as Pluto but on a highly elliptical orbit even farther out which takes 558 years to complete. Its surface is probably made of frozen methane with a little nitrogen and carbon monoxide ice. Haumea and Makemake are smaller icy worlds on distant eccentric orbits beyond Neptune. Dwarf planets have not cleared their neighborhood around their orbit of other similar objects.

The main asteroid belt lies between the orbits of Mars and Jupiter where the gravitational pull of Jupiter prevented planet formation. It contains millions of rocky objects called asteroids that range in size from a few meters to almost 1,000 km across. The three largest asteroids are Ceres, Vesta and Pallas. Asteroids are irregular or elongated in shape since they lack the mass to form spheres. Around half the mass of the asteroid belt is contained in the four largest objects: Ceres, Vesta, Pallas and Hygiea.

Most asteroids formed over 4 billion years ago and orbit the Sun on stable paths, but collisions can send them on erratic trajectories. Some asteroids have moons or orbit in pairs. Asteroids that cross Earth's orbit are called near-Earth objects. Around 2,000 near-

Earth asteroids wider than 1 km are known and monitored. Millions of tiny asteroids under 100 meters wide also exist. Occasionally large asteroids collide with Earth and other planets - an event that may have caused mass extinctions in the distant past. Asteroids are composed of rock, iron and nickel and are leftovers from the formation of the solar system. They give clues about the early solar system since they have undergone minimal geological changes over billions of years. Space probes such as NASA's Dawn have visited and orbited some larger asteroids like Vesta and Ceres for close-up analysis. Future asteroid mining may extract resources like precious metals, water and rocket fuel.

Comets

Comets are icy objects that develop a tail of gas and dust when they are heated near the Sun. They originate from two cold outer regions of the solar system: the Kuiper Belt beyond Neptune and the Oort Cloud extending halfway to the nearest stars. Their composition includes frozen water, methane, ammonia, carbon dioxide, carbon monoxide and other volatiles. Comet nuclei range from a few kilometers to tens of kilometers wide but have extremely low densities. When heated near the Sun, they form an atmosphere called a coma and release dust and gases into a tail that can be thousands or millions of kilometers long.

Short-period comets originate in the Kuiper Belt while long-period comets come from the distant Oort Cloud region. As they pass through the inner solar system, the solar wind and radiation pressure gradually disintegrate the comet's ices. Sungrazing comets dive extremely close to the Sun on their orbits. Famous periodic comets include Halley's Comet, visible every 76 years, and Encke's Comet with a 3-year period. Comet tails always point away from the Sun due to the solar wind. They may have two tails: a dust tail and an ion tail consisting of ionized gases.

Comets may have delivered some of the water and organic compounds that allowed life to form on Earth and other planets.

Asteroid impacts and comet activity like the Late Heavy Bombardment may have brought much of the water in Earth's oceans. Ground and space telescopes continue to discover new comets, both ones passing through the inner solar system for the first time and periodic comets on regular orbits. NASA has carried out close flyby missions to investigate several comets like Halley, Tempel-1, Borrelly and Wild-2 to analyze their composition. Future comet sample return missions are planned to bring back frozen material for laboratory study.

Meteorites

Meteorites are fragments of asteroids, comets or other bodies that survive passage through a planet's atmosphere and collision with the surface. Small examples are known as meteoroid or shooting stars. Larger ones can explode in the atmosphere as bolides or impact the ground as meteorites. Tens of millions of kilograms of extraterrestrial material in the form of micrometeorites and larger meteorites fall on Earth each year.

Most meteorites originate from collisions and fragmentation of asteroids. They provide important clues about the formation and composition of the early solar system since asteroids are relatively unchanged over billions of years. Asteroids samples via meteorites avoid the need for expensive sample return space missions. But meteorites can also come from the Moon and Mars as material ejected by impacts. Intense shock from impacts can alter meteorites, so the most pristine ones are most useful. Rare carbonaceous chondrite meteorites contain organic compounds and water.

Meteorites are categorized into three main types by their structure and composition: stony meteorites or aerolites, which are rocks and account for over 90% of meteorite falls; iron meteorites with iron-nickel alloy content; and stony-iron meteorites that contain both rock and metal phases. Within stony meteorites, chondrites retain the original granular structure while achondrites are igneous rocks crystallized from melt. Meteorite specimens are named based on where they fell. Museums around the world curate meteorite collections.

Monitoring systems like NASA's All Sky Fireball Network use photographic and radar data to triangulate meteorite fall locations. Searching Antarctic ice where meteorites accumulate over time has also yielded many specimens. New meteorite falls are analyzed for information on asteroids and the early solar system. But meteorites have also raised concerns about possible impacts on Earth. The asteroid that created Meteor Crater in Arizona 50,000 years ago was only about 50 meters wide. Far larger meteors produced the 180 km wide Chicxulub Crater linked to dinosaur extinction.

Interesting Quotes about Our Solar System

Our solar system is a wondrous collection of planets, moons, asteroids with a brilliant star at its heart. Needless to say, it has been a source of intrigue and discovery for centuries. The scale and variety of our solar system has captivated the imaginations of many, including some of the most thoughtful minds in history. Here are some insights they've shared:

- "The Sun, with all those planets revolving around it and dependent on it, can still ripen a bunch of grapes as if it had nothing else in the universe to do." - Galileo Galilei
- "The stars in the heavens sing a music if only we had ears to hear." - Pythagoras
- "This most beautiful system [The Universe] could only proceed from the dominion of an intelligent and powerful Being." - Isaac Newton
- "Astronomy compels the soul to look upwards and leads us from this world to another." - Plato
- "To confine our attention to terrestrial matters would be to limit the human spirit." - Ptolemy

- "He who studies the stars will discover an order and harmony perfect and complete, and will realize that all things progress in accordance with a divine plan." - Marcus Manilius
- "We are star stuff which has taken its destiny into its own hands." - Carl Sagan
- "For my confirmation, I didn't get a watch and my first pair of long pants, like most Lutheran boys. I got a telescope. My mother thought it would make the best gift." - Carl Sagan
- "We have lingered long enough on the shores of the cosmological ocean. The time has come to set sail!" - Carl Sagan
- "Every sunset brings the promise of a new dawn." - Ralph Waldo Emerson

These reflections offer a glimpse into the fundamental connection we have with the cosmos and our aspirations to explore and settle beyond our home planet. They remind us of our origins, our potential, and the perpetual cycle of endings and beginnings mirrored in the cosmos itself.

Recent Solar System Discoveries

Our knowledge of the solar system has expanded enormously in recent years thanks to advanced telescopes, planetary probes, and new detection techniques. Some key discoveries about the solar system in the past 5-10 years include:

- NASA's New Horizons probe flew by Pluto and its moons in 2015, revealing Pluto's complex geology with icy mountains and plains, plus a hazy multilayered atmosphere. More recent data indicates Pluto may have a liquid ocean beneath its surface.
- The Juno spacecraft went into orbit around Jupiter in 2016 and has provided unprecedented close-up measurements of its atmosphere, magnetic fields, auroras and internal

structure. Findings show Jupiter's iconic Great Red Spot storm extends hundreds of kilometers down.

- Studies of the solar wind and Mars' atmosphere by probes like MAVEN have given insights into how Mars lost its thicker atmosphere and liquid surface water over billions of years.
- The OSIRIS-REx spacecraft collected a sample from the asteroid Bennu in 2020 which will be returned to Earth in 2023. This will provide scientists with pristine asteroid material to study in the lab.
- New moons continue to be discovered around Jupiter, Saturn and Neptune through telescope observations. Saturn is now known to have over 80 confirmed moons, with dozens more awaiting confirmation.
- Evidence of possible hydrothermal activity and recent volcanic flows has been detected on Venus, indicating it may still be geologically active with an internal heat source.
- Observations of Jupiter's moon Ganymede have found evidence of salt water oceans beneath its icy surface, making it a potentially habitable environment.
- Improved astronomical techniques such as radial velocity measurements, transit photometry and microlensing have led to over 4,000 confirmed exoplanets outside our solar system. Strange new worlds are being discovered yearly.
- Space missions to comets like Tempel-1, Wild-2 and 67P/Churyumov-Gerasimenko have analyzed their composition via close flybys and even return of comet samples. Icy comets contain complex organic compounds and clues to solar system formation.
- Studies of meteorites, lunar samples and Mars rocks continue to teach us about the formation and geology of the early solar system and allow us to date events like asteroid collisions.
- New perturbations in the orbits of extreme trans-Neptunian objects indicate there may be an undiscovered massive planet, sometimes referred to as Planet Nine, in the distant outer solar system.

In summary, a golden age of planetary science has revealed an enormous amount about the worlds in our solar system and how they formed and evolved over 4.5 billion years. Each new discovery builds a more complete picture of the dynamic processes and chemistry that shaped the planets and small bodies around our parent star, helping understand Earth's own origins. Exciting missions to ocean worlds, asteroids and the gas giants lie ahead. The ancient question "Are we alone?" also continues to draw us outward, searching for life beyond Earth.

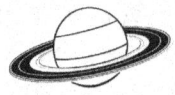

Chapter 2: Stars and Constellations

Types of Stars

Stars are enormous spheres of plasma that generate heat and light through nuclear fusion in their cores. The closest star to our solar system is the Sun, which is classified as a G-type yellow main sequence star. There are estimated to be over 100 billion stars in our Milky Way galaxy and even more galaxies in the observable universe.

Stars come in a range of sizes, temperatures, luminosities, and colors. The smallest stars are red dwarfs, followed by yellow dwarfs like our Sun. Intermediate size stars are subgiants and bright giants. The largest stars are super giants and hypergiants. After a star has fused all its nuclear fuel, its core collapses and the star sheds its outer layers to become a white dwarf, neutron star, or black hole.

Red dwarfs comprise about 70% of the stars in the Milky Way. They have the coolest surface temperatures, from 2400K to 3800K, giving them a reddish hue. Although small in physical size, red dwarfs can burn for trillions of years due to their low luminosity and slow rate of fusion. Proxima Centauri, the nearest star at 4.2 light-years from Earth, is a red dwarf. Some have intense flares that may sterilize any planets.

Yellow dwarfs make up about 10% of stars. This main sequence includes the Sun with a surface temperature around 5800K. Yellow dwarfs convert hydrogen into helium via the proton-proton chain reaction in their cores for 10 billion years on average. Besides the Sun, Sirius A, Tau Ceti, and Epsilon Indi are examples. They are several times larger than red dwarfs and emit much more light. Blue-white dwarf stars have the hottest surface temperatures of 10,000K to 50,000K. They include the blue-white companion star Sirius B, a white dwarf with a mass equal to the Sun but a size similar to Earth. Due to their high mass, blue-white dwarfs have short lifespans of a few hundred million years. Rigel, Deneb, and many other blue supergiants are also in this category before becoming supernovae.

Betelgeuse is a well-known cool red supergiant star. Its diameter is about 1000 times that of the Sun, larger even than Jupiter's orbit. Betelgeuse has a surface temperature of only 3600K but is a staggering 100,000 times more luminous than the Sun due to its immense size. Antares and VY Canis Majoris are other red supergiants. Despite their great size, red supergiants only live for 10 - 50 million years.

At the top end, hypergiants like Rigel have over 30 times the mass of the Sun and are among the most luminous stars known. They emit hundreds of thousands of times the luminosity of the Sun and are very unstable, blowing off massive shells of plasma and gas. Their lifetimes are just a few million years before going supernova. After a star has completed fusion of elements in its core, it begins to die. Its core collapses and the star sheds its outer layers. Small and medium stars become white dwarfs, which are extremely dense cores of carbon and oxygen atoms surrounded by a thin atmosphere. Sirius B and Procyon B are examples. If the star had enough mass, further collapse yields a neutron star. Even more massive stars end their lives as black holes.

Constellations and Asterisms

The night sky has fascinated humans since antiquity. Ancient cultures saw patterns in the stars and perceived them as pictograms of people, animals, and objects, using them to tell stories to remember and pass down through generations.

Constellations are defined areas of the sky containing prominent visible stars that form an imaginary outline or shape. The Ancient Greeks first catalogued over 40 constellations based on legends of their mythology. The Romans later added and formalized constellations using the zodiac along the ecliptic plane. Other cultures also invented constellation patterns, including the Inuit of the Arctic and Aboriginal Australians.

In modern times, the International Astronomical Union (IAU) officially recognizes 88 constellations that divide up coverage of the entire celestial sphere. The constellations vary greatly in physical size and shape. The largest is Hydra that spans over 1300 square degrees. The Southern Cross or Crux constellation by contrast covers only 68 square degrees.

The IAU constellations contain all of the individual stars visible to the naked eye from Earth. Well known northern constellations include Ursa Major, Cassiopeia, Orion the Hunter, Leo the Lion, and Draco the Dragon. Orion is one of the most prominent due to its hourglass shape, nearby bright stars like Rigel and Betelgeuse, and Orion's Belt of three stars in a row.

Notable southern constellations visible from temperate latitudes include Crux the Southern Cross, Centaurus the Centaur, Carina the Keel, and Canis Major which contains Sirius, the brightest star. Crux points toward the southern celestial pole and is depicted on the flags of many southern nations.

Constellations should not be confused with asterisms like the Big Dipper and the Little Dipper. Asterisms are prominent and recognizable patterns of stars that are within a larger constellation. The Big Dipper and Little Dipper are part of Ursa Major and Ursa Minor but are not constellations in their own right. Other examples include the Summer Triangle of three bright stars from different constellations and the Winter Hexagon of seven stars that include Rigel and Sirius.

Since stars slowly move relative to each other over millennia, the constellations gradually change their shapes. However, their legacy as groupings of stars that guided ancient mariners and explorers remains culturally significant. Long-exposure photographs reveal the spectacular star trails as Earth rotates, forever circling the constellations.

Interesting Facts About Stars

- The nearest star to our solar system is Proxima Centauri at 4.2 light years away. The next closest are Alpha Centauri A and B, forming a triple system with Proxima. Over 270,000 stars are estimated to lie within 50 light years of Earth.
- Betelgeuse, a red supergiant star located at the shoulder of Orion, is one of the largest stars known. Its diameter is estimated to be over 1300 times that of our Sun, larger even than the orbit of Jupiter.
- Blue hypergiant star Rigel in Orion is over 120 times more massive than the Sun and shines with the luminosity of over 120,000 Suns. It also has a radius nearly 80 times larger than the Sun.
- Red dwarf TRAPPIST-1 has seven Earth-sized exoplanets orbiting it, with at least three in the habitable zone at the right temperature for liquid water to potentially exist on their surfaces.
- Eta Carinae, one of the most massive and luminous stars observed, had a massive explosion in 1843 that was the second brightest supernova after the Crab Nebula

supernova. This giant eruption hurled material into space that still forms the Homunculus Nebula seen around Eta Carinae.

- Neutron star J1748-2446ad located near the center of the Milky Way spins over 700 times a second, making it the fastest spinning celestial object ever recorded. Some theories suggest it may be a rapidly spinning quark star rather than a neutron star.
- Kepler-11 is a Sun-like star located 2000 light years away that has six confirmed exoplanets, five in tight orbits closer than Mercury while the sixth orbits where Earth is located. This Kepler-11 system resembles a miniature version of our inner solar system.
- Lalande 21185 is one of the nearest stars to the Sun at just 8.3 light years away. It is too faint to see with the naked eye at seventh magnitude but was the first star found to have a planetary system with two confirmed exoplanets.
- Sirius is the brightest star visible from Earth, twice as luminous as the next brightest star Canopus. Sirius A is a white main sequence star while its companion Sirius B is a super dense white dwarf, the leftover core of a dead star.
- The Coalsack Nebula near the Southern Cross is the most prominent dark nebula visible to the naked eye, blocking the light from more distant stars. It has a size of about 5 degrees and contains enough dust to form over 10,000 Sun-like stars if it collapsed.

Noteworthy Quotes About Stars and Constellations

Stars and constellations have been subjects of human fascination for millennia, serving as the backdrop for our greatest stories and the source of our most fundamental substances. Let us ponder some simple yet profound thoughts about these celestial bodies. In this section, we focus exclusively on the quotes of Carl Sagan. Sagan was a renowned astronomer and a masterful communicator who had an exceptional ability to translate the complexities of the cosmos into language that resonated with people from all walks of life. His passion for the stars and his eloquent way of sharing that enthusiasm made the vast universe seem accessible and intimately connected to our everyday lives. Through his books, television appearances, and especially his groundbreaking series "Cosmos," Sagan invited us all to look at the sky not just with wonder, but with understanding. He was not only a scientist but also a storyteller, one who painted the narrative of our universe in strokes that captivated the imagination of millions around the world.

- "The nitrogen in our DNA, the calcium in our teeth, the iron in our blood, the carbon in our apple pies were made in the interiors of collapsing stars. We are made of starstuff." — Carl Sagan
- "We are a way for the cosmos to know itself." — Carl Sagan
- "Somewhere, something incredible is waiting to be known." — Carl Sagan
- "For small creatures such as we the vastness is bearable only through love." — Carl Sagan
- "The cosmos is full beyond measure of elegant truths; of exquisite interrelationships; of the awesome machinery of nature." — Carl Sagan

In essence, these thoughts bring us closer to the realization that we are intimately linked with the cosmos. From the elements that form our bodies to the curiosity that propels our minds, we are a part of the universe as much as it is a part of us. The stars above us are not just points of light but are the very fabric of our being, and in understanding them, we understand ourselves. As we continue to look up and learn, we find that love and discovery make the immensity of the universe not just bearable, but wondrous.

Recent Developments in Stellar Astronomy

Some key developments in the study of stars and stellar systems in the past decade include:

- The Gaia space observatory launched in 2013 has mapped the precise 3D positions, distances and motions of over 1 billion stars within the Milky Way galaxy. This is allowing very detailed mapping of our galaxy.
- TESS, the Transiting Exoplanet Survey Satellite launched in 2018, has discovered over 200 confirmed exoplanets orbiting nearby bright stars which are suitable for follow-up atmospheric analysis.
- ALMA, the Atacama Large Millimeter Array in Chile that began full operations in 2013, obtains extremely detailed radio and submillimeter images of galaxies, star formation regions, protoplanetary systems, and more.
- LIGO and Virgo have detected ripples in spacetime called gravitational waves from collisions of dense dead stars like black holes and neutron stars. This provides a new window into these events.
- New space telescopes like TESS, Gaia, NICER, and eROSITA use near-infrared, X-ray and other wavelengths to peer through dust clouds and analyze stars forming inside nebulae.
- Detailed models of nuclear fusion inside different types of stars have improved our understanding of how elements are synthesized and dispersed into space when the largest stars explode.
- Astronomers found a hypervelocity star was ejected from the center of the Milky Way galaxy likely by interacting closely with the supermassive black hole Sagittarius A*.
- Studies of extremely metal-poor stars indicate the first stars formed only 250 to 350 million years after the Big Bang during the Cosmic Dark Ages before galaxies existed.
- Observations show some massive star binaries can explode as supernovae when their partner star strips away their outer layers, challenging models of how they evolve.

In summary, astronomy has made remarkable progress recently in unraveling the mysteries surrounding the stellar engines that power galaxies and create the chemical elements. The study of stars, nebulae, and exoplanets will continue to reveal ever more about the origins of cosmic structures and life in the universe.

Chapter 3: Galaxies and Cosmology

Introduction

Our home galaxy, The Milky Way, contains over 200 billion stars, including our Sun, all orbiting out in one spiral arm. Yet our galaxy itself spans only around 100,000 lightyears across – a relative speck among hundreds of billions of other galaxies dotting the observable universe. The study of galaxies and their origins and evolution provides a window into the grand story of cosmic history since the Big Bang started it all 13.8 billion years ago. Let us start by considering the types of galaxies that populate our solar system and universe.

Types of Galaxies

Galaxies are colossal assemblies of stars, gas, dust and dark matter bound together by gravity. They can contain anywhere from a few million to over a trillion stars along with interstellar clouds of gas and dust. The Milky Way galaxy that houses our solar system has around 200-400 billion stars.

Galaxies exhibit a variety of shapes and structures that are classified into four main types: spiral, elliptical, irregular, and lenticular. Their distinct shapes reflect differences in stellar orbits, rotational dynamics, and evolutionary history. Many galaxies also have smaller satellite dwarf galaxies in their vicinity.

Spiral galaxies like our Milky Way account for over 70% of bright galaxies. They have central bulges surrounded by flattened rotating disks of stars, gas and dust with distinctive spiral arms that emerge from the nucleus and wrap around the galaxy. Spiral arms contain young, hot blue stars and nebulae where new stars are forming. Elliptical galaxies range from nearly spherical to very elongated ovals but lack any spiral features. They contain mostly old stars loosely distributed in a spheroidal shape with radial orbits around the center. Ellipticals can be nearly gas and dust free. Giant elliptical galaxies are the largest and most massive type.

Irregular galaxies have no organized spiral or elliptical structure. Their stars follow random disordered orbits. Irregulars are rich in gas and dust and undergo intense star formation activity. Their chaotic shapes are likely the result of gravitational interactions with other galaxies. The Magellanic Clouds orbiting the Milky Way are irregular galaxies.

Lenticular galaxies represent a transition type. They have a disk shape similar to large spirals but lack any spiral arms and contain mostly old stars. Dust lanes are also absent. They are most often found near large ellipticals and may represent spiral galaxies that have used up their gas.

Dwarf galaxies have less than about 10 million stars. Despite their small size, they are actually the most common galaxy type, outnumbering giant galaxies. Dwarf spheroidal galaxies orbit larger galaxies like the Milky Way as satellite galaxies. Other irregularly shaped dwarfs include the Large and Small Magellanic Clouds.

The Milky Way Galaxy

The Milky Way is our home spiral galaxy and part of the Local Group of galaxies. From Earth, it appears as a diffuse band of white light arcing across the night sky. This luminous band is caused by the combined light of over 200 billion stars and associated dust in the galactic plane viewed edge-on from our position inside the disk. The Milky Way has a diameter of 100,000 to 200,000 light-years.

The galactic center at the core of the Milky Way has extremely dense stellar crowds. Observations of stars orbiting the core have revealed a supermassive black hole 4 million times the mass of the Sun, named Sagittarius A*. Surrounding it is a nuclear stellar cluster with stars packed only light-days apart.

Spiraling out from the central hub, the Milky Way disk has a radial size of 50,000 to 60,000 light-years and only 1000 light-years thick. It contains the spiral arms that give the galaxy its pinwheel structure made of hydrogen gas, dust, and young stars. The solar system resides around 26,000 light-years from the galactic center, within the Orion Arm segment between the Sagittarius and Perseus Arms.

Beyond the stellar disk, the Milky Way has an extended roughly spherical halo component 100,000 to 200,000 light-years in diameter. The halo is dominated by the galactic corona of hot tenuous gas along with globular clusters and extremely old stars. And permeating all parts of the Milky Way is invisible dark matter that provides most of the galaxy's mass.

Dark Matter and Dark Energy

In the last century, astronomers discovered two mysterious components of our universe - dark matter and dark energy - that remain unexplained but dominate the cosmos.

Dark matter is an invisible substance that accounts for 85% of all matter in the universe. Its existence is inferred from unexplained mass discrepancies in galaxies, gravitational lensing and distortions of light paths, and measurements of cosmic microwave background fluctuations and structure formation. Without dark matter providing extra gravity, galaxies would fly apart from their rapid rotations. But the particle nature of dark matter remains unknown.

Dark energy is an unknown form of energy permeating all of space and accelerating the expansion of the universe. It represents about 68% of the total energy density of the universe. The discovery of dark energy in the late 1990s upended cosmology, indicating empty space has inherent energy. Understanding dark energy may require new physics beyond general relativity, possibly a dynamical field like quintessence or a cosmological constant related to vacuum energy and quantum theory.

Together, dark matter and dark energy point to immense gaps in our fundamental understanding of particles, fields, and the cosmos. Their discovery forced a major revision in the standard big bang cosmological model. Observations continue to precisely measure their occasionally contradictory effects to narrow down theoretical possibilities. Unraveling the mysteries of dark matter and dark energy is one of the greatest challenges in physics and astronomy.

Quotes About Galaxies and the Universe

The universe is a complex tapestry of cosmic events and entities that often defy human understanding, reminding us that there are truths out there which may not align with our natural intuitions. The following quotes from various influential figures offer a perspective on this enigmatic cosmos that surrounds us:

- "The universe is under no obligation to make sense to you."
  - Neil deGrasse Tyson
- "We spend too much time staring into glowing rectangles and not enough time gazing into the starry sky." - Michelle Obama
- "The universe seems neither benign nor hostile, merely indifferent." - Carl Sagan
- "Some part of our being knows this is where we came from. We long to return. And we can. Because the cosmos is also within us. We're made of star stuff. We are a way for the cosmos to know itself." - Carl Sagan
- "The nitrogen in our DNA, the calcium in our teeth, the iron in our blood, the carbon in our apple pies were made in the interiors of collapsing stars. We are made of starstuff." - Carl Sagan
- "Astronomy compels the soul to look upward, and leads us from this world to another." - Plato

These quotes underscore a profound truth: our relationship with the universe is intrinsic and enduring. The pursuit to understand it is not just an academic exercise but a reflection of our very nature. The universe, with its indifference and mystery, invites us to learn, to explore, and to appreciate the intricate dance of existence in which we are all participants.

New Observations in Cosmology

Some key findings in cosmology and galaxy studies in the past decade include:

- Detailed observations of the cosmic microwave background by the Planck spacecraft further constrained models of the early universe and parameters like its geometry, content, and expansion rate.
- The Dark Energy Survey precisely measured gravitational lensing, galaxy distributions and supernovae to map the effects of mysterious dark energy that is accelerating cosmic expansion.

- Detection of gravitational waves from colliding neutron stars allowed novel tests of general relativity and provided a new independent way to measure the rate of the universe's expansion.
- The ALMA and Hubble telescopes have observed early galaxies rapidly forming soon after the Big Bang, revealing details about cosmic reionization and structure growth.
- Studies found supermassive black holes at the centers of most large galaxies grew and became active much earlier than expected, as far back as when the universe was only 5% its current age.
- Multiple analyses revealed the Andromeda Galaxy is on a collision course with our Milky Way, likely merging in a spectacular cosmic collision in approximately 4 billion years.
- Astronomers discovered our home supercluster of over 100,000 galaxies, named Laniakea, is being gravitationally pulled by the massive Great Attractor region 250 million light-years away.
- New techniques for mapping stellar streams originating from disrupted galaxies provided insights into the detailed formation history of our Milky Way galaxy and its past mergers.
- Observations indicate dark matter may be "warmer" and have greater velocities than predicted by the standard cold dark matter model, suggesting it interacts and scatters weakly.

Understanding dark energy, cosmic inflation, the composition of the universe, and the growth of structure remains key areas of research in cosmology. Upcoming telescopes like the James Webb Space Telescope will peer deeper into the observable universe than ever before. The cosmos beckons.

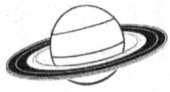

Chapter 4: Black Holes and Wormholes

Understanding Black Holes

Black holes are extraordinarily dense regions of spacetime exhibiting such strong gravity that nothing can escape from inside them, not even light. They form from the collapsed cores of massive stars or direct collapse of dense clusters of matter. Supermassive black holes with masses millions to billions of times that of the Sun exist at the centers of most large galaxies including our own Milky Way galaxy.

When massive stars with at least 20-30 solar masses exhaust their nuclear fuel, they explode as powerful supernovae and their cores implode. If this stellar remnant core exceeds about 3 solar masses, the enormous gravity crushes it further into a singularity with infinite density – a black hole. This leads to extreme warping of spacetime around the singularity. Black holes therefore represent a breakdown of known physics at their central points.

The boundary around every black hole where the escape velocity equals the speed of light is called the event horizon. Passing through this horizon means no return; time appears frozen for a falling observer aiming toward the center. However, to outside observers, it takes an infinite amount of time for infalling matter to cross the event horizon, making it fade away and appear to asymptotically approach the border, thanks to relativistic time dilation effects.

Black holes continue growing and pull in gas, dust and stars from their surroundings into an accretion disk circling the event horizon. As matter accelerates toward the boundary, collisions cause extreme friction heating up the infalling material to millions of degrees, making the disks glow intensely with high-energy radiation like X-rays. Supermassive black holes can consume millions of Solar masses worth of material via accretion over hundreds of millions of years.

When pairs of black holes spiral around each other closely, they emit gravitational waves that propagate through spacetime. In 2016, the LIGO-Virgo collaboration detected these ripples for the first time coming from two ~30 solar mass black holes that merged into one bigger black hole in a fraction of a second. Dozens more stellar mass black hole collisions have since been observed. Supermassive black hole pairs at galactic centers may eventually merge as well, sending even more energetic gravitational waves sweeping across the universe.

Rotating black holes have two additional key surfaces - the inner and outer event horizons. The ergosphere boundary outside the horizon indicates space is dragged around by the spin. Matter entering this region can gain energy from the rotation. Electrons crossing the inner horizon at close to lightspeed get crushed and emit gamma rays in powerful bursts. When absorbing matter, some jets at the black hole's poles can eject a small fraction back into space at nearly light speed. Quasars are examples of highly active galactic nuclei thought powered by supermassive black holes consuming gas and stars.

Wormholes and Time Travel

Wormholes are theoretical tubes or passages through the fabric of spacetime that could connect distant points in the universe or even different universes. While no natural stable wormholes have ever been discovered, they are predicted as a valid mathematical solution to the field equations of Einstein's general theory of relativity. Hypothetical traversable wormholes would act like shortcuts or tunnels through the curvature of spacetime.

Some theories describe wormholes potentially joining extremely long distances perhaps billions of light years apart, massively reducing travel time to distant stars or galaxies. Another intriguing but highly speculative idea is that wormholes could link other universes or realities, functioning as portals to parallel dimensions according to a cosmology of the multiverse containing infinite branched universes. However, actual existence, stability, and traversability of wormholes remain open questions.

If wormholes are real and could be stabilized and enlarged to allow passage, they would offer the possibility of time travel by providing a shortcut between two separate points in space and time. For example, a wormhole could connect Earth in 2050 with Alpha Centauri in 2150. The two mouths of the wormhole experience different passage of time. Carefully timed transit through the wormhole could allow moving into the future or past.

Attaining sufficient negative energy density to maintain wormhole dimensions open against collapse poses one of the largest challenges for forming traversable wormholes. Quantum effects might be used to hold a wormhole open but could only transmit information, not matter. For macroscopic wormholes, exotic matter with anti-gravity effects would be required. The energy requirements appear immense given our current physics knowledge.

While traversable wormholes remain theoretical, better understanding the quantum structure of spacetime and effects like quantum entanglement may provide insights about information transmission through hyperspace. Space-time distortions created by supermassive black hole pairs could also potentially be used to engineer stable wormholes. But practical application of wormholes for any form of matter transmission or time travel appears highly unlikely given immense technical limitations. For now, wormholes feature only in speculative science fiction rather than reality.

Quotes About Black Holes and Wormholes

Black holes, the enigmatic titans of space, have long captured the human imagination and posed a multitude of questions for scientists and thinkers alike. They represent regions of space where the known laws of physics converge into a question mark. These cosmic phenomena bend the fabric of space and time to such an extent that our current understanding strains to comprehend their full nature. The following quotes from eminent minds offer a glimpse into the awe-inspiring and sometimes baffling nature of black holes:

- "Black holes are where God divided by zero." - Albert Einstein
- "Even light, which travels so fast it circumnavigates the Earth 8 times in one second, will take about 100,000 years to travel from one side of a black hole to the other." - Michio Kaku, physicist
- "A wormhole would allow us to travel from one side of the galaxy to another side of the galaxy almost instantaneously." - Ronald Mallett, physicist
- "Black holes are the seductive dragons of the universe." - Neil deGrasse Tyson, astrophysicist
- "We all exist within space-time, which itself looks empty. But the worms of black holes prove it ain't so." - John Wheeler, physicist

As we stand on the precipice of new astronomical eras, with technology inching us ever closer to unlocking the mysteries of the universe, black holes remain as both a challenge and a beacon. They beckon us to delve deeper, push further, and think more profoundly about the cosmos. These celestial enigmas are not just puzzles to be solved but are also metaphors for the vastness of the unknown that lies before us. They symbolize the continuous quest for knowledge and understanding in the grand expanse of the universe that is both humbling and exhilarating. As we expand the frontiers of science and peer into the abyss of black holes, we inch closer to unveiling the immense secrets and uncharted territories that await our future understanding.

Recent Black Hole Discoveries

- First image of a black hole's event horizontaken by the Event Horizon Telescope of the supermassive black hole at the center of galaxy M87.
- LIGO and Virgo gravitational wave detectors have observed many black hole-black hole mergers indicating abundance of "stellar mass" black holes.
- Observations show supermassive black holes at galactic centers undergo periods of intense flare-up from consuming material.
- Analyses reveal some giant black holes are offset from the galactic centers due to gravitational wave recoil from black hole mergers.
- Studies found "intermediate mass" black holes in the 100 - 100,000 solar mass range exist from runaway growth of massive stars.
-

Wormhole Research

- Quantum entanglement between pairs of photons hints that quantum wormholes may be possible.
- Exotic matter with hypothetical negative mass is proposed to stabilize wormholes enough for traversal.
- Mathematical models investigate the possibility that inflationary forces in the early universe could have created primordial wormholes.
- Simulations of extremely dense neutron stars collapsing into black holes suggest tunnels through spacetime may briefly form.
- Physicists calculate what properties traversable wormholes would require to avoid paradoxes like time travel.
- Thought experiments conceive ways advanced civilizations may theoretically construct and maintain stable wormholes.
- No convincing evidence for natural or artificial wormholes has been verified. Fundamental obstacles around exotic matter and paradoxes remain.

Chapter 5: Exoplanets and the Search for Life

"Light thinks it travels faster than anything but it is wrong. No matter how fast light travels, it finds the darkness has always got there first, and is waiting for it." - Terry Pratchett

Overview of Exoplanets

Exoplanets are planets that orbit stars outside our solar system. Thousands have been discovered in the last 25 years using advanced techniques, indicating planets are common in the Milky Way galaxy. Exoplanets can range from small rocky terrestrial planets like Earth to huge gas giants significantly larger than Jupiter. Many star systems have been found with multiple exoplanets orbiting their host stars. Various methods are used to detect exoplanets, including looking for transits as they pass in front of their star from our viewpoint, measuring the gravitational wobble their orbits cause in stars, gravitational microlensing effects, and direct imaging.

Exoplanets exhibit a wide range of characteristics unlike the worlds in our own solar system. Super-Earths are rocky planets bigger and more massive than Earth but smaller than ice giants like Neptune and Uranus. Hot Jupiters are gas giants orbiting extremely close to their parent star. Mini-Neptunes are medium-sized worlds surrounded by thick atmospheres of light elements. Analyzing exoplanetary properties provides clues about how planetary systems form, evolve, migrate, and interact early in their development in protoplanetary disks. The search continues for exoplanets most similar to Earth that may have conditions suitable to support alien life.

Conditions for Habitability

For an exoplanet to potentially harbor life, it must meet conditions to be habitable by having the ability to sustain liquid water on its surface over significant geological periods. The habitability of a world depends on key factors like its distance from the host star, atmospheric composition, and surface temperature. Stars like our Sun provide frequent targets in exoplanet searches focused on a habitable zone at ideal distances for life-bearing worlds.

Main sequence G dwarf stars with long stable lifetimes offer prime candidates around which habitable zone planets could evolve over billions of years. Red dwarf M stars are most abundant but can be too dim and have too long pre-main sequence phases unsuitable for life unless planets orbit very closely. For habitable surface conditions, an exoplanet should have atmospheric pressure and greenhouse gases like carbon dioxide, methane and water vapor to insulate and regulate temperatures.

Additional criteria are considered when evaluating potentially habitable exoplanet candidates, including their size, mass, density, temperatures, atmospheric loss rates, and potential for geologic activity. Best candidates are planets with active geology but not overactive volcanic activity or extreme plate tectonics that disrupt evolved complex biospheres. Life likely requires renewed sources of chemical energy over time and protection from harsh space and stellar radiation.

Quotes About Alien Life

The concept of alien life stirs the imagination and provokes a multitude of questions about our place in the universe. The following quotes reflect the intrigue and curiosity that the possibility of extraterrestrial beings inspires in scientists, writers, and thinkers alike:

- "The universe is a pretty big place. If it's just us, seems like an awful waste of space." - Carl Sagan
- "If there is one thing the history of evolution on Earth has taught us it's that life will not be contained." - Michael Crichton
- "Perhaps we've never been visited by aliens because they have looked upon Earth and decided there's no sign of intelligent life." - Neil deGrasse Tyson
- "It's not a matter of if, but when we'll find alien life." - Michio Kaku
- "The discovery of extraterrestrial life, perhaps more than any other single event, would alter our concept of who we are and where we stand in the cosmic scheme of things." - James Trefil
- "Not to search for life elsewhere would be a failure of human curiosity." - Reza Aslan
- "It is possible that the future of human civilization depends on the ethical treatment of extraterrestrial life forms." - Nick Bostrom
- "The high probability of the existence of life outside our planet makes the search for extraterrestrial intelligence one that we must pursue." - Carol Stoker
- "If we ever encounter extraterrestrial intelligence, I believe it is overwhelmingly likely to be post-biological in nature." - Paul Davies
- "Aliens might be staring at us in disbelief that we cannot see them." - Seth Shostak
- "The thought of being the only intelligent species in a dumb universe is the second worst existential nightmare.

The worst is being watched by an intelligence greater than our own." - Charles Stross
- "Life is a miracle, and it doesn't have to be confined to Earth-like conditions to exist." - Avi Loeb

These quotes express a range of perspectives on the search for and potential impact of discovering life beyond Earth. They reflect a mix of optimism, caution, humor, and profound contemplation about what such a discovery would mean for humanity's understanding of life itself and our place in the cosmos.

Recent Exoplanet Discoveries

- NASA's Kepler space telescope mission identified over 2,600 confirmed exoplanets along with thousands of additional candidates from observing segments of the night sky.
- The Transiting Exoplanet Survey Satellite (TESS) has found hundreds of exoplanets transiting nearby bright stars, providing prime targets for further detailed study and characterization.
- Ground-based searches using precise radial velocity measurements continue to unveil exoplanets by the wobble they induce in their star's motion from gravitational tugs.
- New space telescopes like CHEOPS and the upcoming James Webb are designed to precisely analyze exoplanet atmospheres for molecular signatures and conditions indicative of habitability.
- Improved observations indicate approximately 1 in 5 Sun-like stars have an Earth-sized planet orbiting in their habitable zones at distances for liquid surface water.
- Some recent potentially habitable discoveries include the exoplanets K2-18b, pi Mensae c, TOI-700d and LHS 1140 b, all orbiting M dwarf stars and possibly having liquid water under their surfaces.

Ongoing SETI Research

The search for extraterrestrial intelligence (SETI) involves experiments attempting to detect signs of advanced alien civilizations through interstellar radio signals, optical flashes, probes, spacecraft, and other means.

- Optical and radio telescopes like the Allen Telescope Array and facilities used by the Breakthrough Listen initiative monitor millions of stars and scan large swaths of the sky seeking modulated signals or energy flashes that could indicate technology.
- New directed messaging efforts like METI (Messaging Extraterrestrial Intelligence) have transmitted messages and images like the Arecibo Message via radio telescopes out towards promising star systems to attempt establishing contact.
- Advances in data analysis and artificial intelligence use machine learning algorithms to search enormous sets of radio telescope data seeking anomalous patterns that could represent potential alien communications.
- No definitive evidence of extraterrestrial civilizations through signals or communication has been found in searches so far. But SETI research remains ongoing and holds immense discovery potential if contact is ever made in the vast cosmos.
- Some scientists caution that replying to unknown signals or advertising human presence widely could be risky before better understanding the motivation and nature of alien intelligences.

The possibility of life beyond Earth captivates human imagination and scientific curiosity alike. Exoplanet statistics indicate billions of potentially habitable worlds likely exist in the Milky Way galaxy alone. Future telescopes will thoroughly probe their atmospheres and surfaces. We may stand on the verge of uncovering signs of life beyond our solar system.

Chapter 6: Rocketry and Spaceflight

Introduction

The genesis of space exploration is rooted deeply in the early 20th century when pioneers like Konstantin Tsiolkovsky laid down the theoretical groundwork for astronautics, including rocketry and the potential of multi-stage launch vehicles. This era witnessed the practical application of these theories by Robert Goddard, who successfully launched the first liquid-fueled rocket, and later by Wernher von Braun, whose work was instrumental in the development of the V-2 rocket and eventually the Saturn V, which carried humans to the Moon.

The launching of Sputnik 1 by the Soviet Union in 1957 marked the beginning of space exploration, forever changing the dynamics of international politics and scientific pursuit. This watershed event was closely followed by another monumental achievement when Yuri Gagarin orbited the Earth in 1961, claiming the title of the first human in space. In a swift response, the United States sent Alan Shepard into space aboard the Freedom 7 capsule in a suborbital flight that signified the burgeoning competition in space technology.

These early ventures into the unknown tested not just the resilience of humans in the harsh conditions of space but also the reliability of technology beyond Earth's atmosphere. The experiments conducted during these missions yielded crucial data about living organisms' responses to microgravity and the viability of communications from beyond Earth's envelope.

The Soviet and American space programs served as the vanguard for our venture into space, challenging each other in a spirited rivalry that became known as the Space Race. This period of intense competition spurred rapid developments in a wide range of technological areas and set humanity on a path toward continuous space exploration.

The Space Race

The Space Race was more than a contest of technological prowess; it was an emblem of geopolitical tensions and the ideological contest between the US and USSR that characterized the Cold War. The stakes of the Space Race were not limited to the advancement of space technology but extended to the military, scientific, and even societal domains where dominance in space equated to international prestige and influence.

The Soviet Union secured a series of early victories in this celestial competition, including the first human in orbit and the first probe to touch the Moon's surface. The US, feeling the pressure of these achievements, concentrated efforts on the Mercury, Gemini, and Apollo programs. Project Gemini focused on advanced space travel techniques such as spacewalking and orbital rendezvous, which were prerequisites for the Apollo missions. Apollo's crowning glory was the lunar landing in 1969, a moment that united the world in awe.

Concurrently, the USSR's Vostok and Soyuz programs broke barriers in long-duration spaceflight, and these spacecraft remain operational in various updated forms, underscoring their enduring design. The race also precipitated a cascade of technological innovations, from advanced computing techniques to new materials that have since filtered into various aspects of daily life.

Quotes About Space Exploration

The narrative of space exploration is rich with the indomitable spirit of human curiosity and the desire to push the boundaries of our world. The quotations from pioneers and visionaries in this domain capture the essence of this journey, each reflecting a facet of humanity's enduring quest to explore the unknown.

- "That's one small step for man, one giant leap for mankind." - Neil Armstrong
- "Space exploration is a force of nature unto itself that no other force in society can rival." - Neil Armstrong
- "Mystery creates wonder and wonder is the basis of man's desire to understand." - Neil Armstrong
- "To look out at this kind of creation out here and not believe in God is to me impossible." - John Glenn
- "We can't help it. Life looks for life." - Carl Sagan
- "If we die, we want people to accept it. We are in a risky business, and we hope that if anything happens to us, it will not delay the program. The conquest of space is worth the risk of life." - Gus Grissom
- "There is perhaps no better a demonstration of the folly of human conceits than this distant image of our tiny world." - Carl Sagan, on the Pale Blue Dot photograph
- "I don't think the human race will survive the next thousand years, unless we spread into space." - Stephen Hawking
- "It suddenly struck me that that tiny pea, pretty and blue, was the Earth. I put up my thumb and shut one eye, and my

thumb blotted out the planet Earth. I didn't feel like a giant. I felt very, very small." - Neil Armstrong

These reflective words, stemming from personal experience and introspection, are not simply phrases but embody the dreams, wisdom, and the forward-thinking vision necessary for the continuous exploration of space. They remind us of the courage it takes to step into the unknown and the transformative power of setting foot on uncharted worlds. Through these voices, we can discern the ongoing narrative of our cosmic journey and the imperative to look beyond our earthly confines towards the broader universe.

Recent Space Missions

The current landscape of space exploration is a vibrant tableau of missions that extend our sensorium to the cosmos. NASA's rovers, including Curiosity, Perseverance, and InSight, are more than just machines on Mars; they are our avatars on an alien world, seeking signs of past life and unraveling the mysteries of Martian geology. Their every discovery feeds into our understanding of habitability beyond Earth.

SpaceX has not only realized the dream of reusable rockets but has also successfully transported astronauts to the International Space Station with the Crew Dragon spacecraft. This represents a significant milestone in commercial spaceflight, indicating a future where space travel becomes more routine.

China's Chang'e program has not just landed on the Moon; it has explored the far side, an unprecedented feat that expands our lunar knowledge frontier. The OSIRIS-REx and Hayabusa2 missions have reached out to the ancient rocks of our solar system, touching asteroids Bennu and Ryugu, respectively, and bringing pieces of these primordial bodies back to Earth, providing invaluable insight into our cosmic origins.

India has also emerged as a major player in the global space race, with its ambitious Chandrayaan missions underscoring the country's growing prowess. Chandrayaan, India's lunar exploration program, has achieved significant milestones, including the Chandrayaan-2 mission, which aimed to explore the uncharted lunar south pole. Despite the hard landing of its Vikram lander, the orbiter continues to gather crucial data, contributing to our understanding of the Moon's surface and the potential presence of water ice. You can read more about this mission in my book India's Moonshot.

India's space agency, the Indian Space Research Organization (ISRO), has not only focused on lunar expeditions but has also set its sights on interplanetary journeys. With the successful Mars Orbiter Mission (Mangalyaan), India became the first nation to reach Mars orbit on its maiden attempt, a testament to its technical acumen and cost-effective approach to space exploration. These forays into the unknown reflect a broader vision of harnessing space technology for national development, while also seeking answers to fundamental questions about the universe.

Future Space Exploration Goals

The future of space exploration is bright with ambitious goals. NASA's Artemis I mission serves as a prologue to a renewed era of lunar exploration, setting the stage for a deeper human presence on the Moon. The Artemis program is more than a series of flights; it is the beginning of a sustained human engagement with the lunar environment, potentially establishing a permanent base, with the south pole as a candidate due to its water ice deposits.

SpaceX's Starship and Super Heavy booster are not just vehicles but the harbingers of interplanetary travel, with Mars squarely in their sights. The development of these rockets represents a leap forward in our capabilities to venture beyond Earth-Moon system.

The planned lunar Gateway is more than a space station; it is a portal to the wider solar system, a staging post that will enable unprecedented exploration and serve as a testbed for the technologies required for deeper space exploration.

The quest for signs of life on Mars continues, with the search for biosignatures driving multiple missions. The potential discovery of life, past or present, on Mars would be a profound moment in human history, reshaping our understanding of life's uniqueness and prevalence in the universe.

In pursuing these ambitious goals, we are not just exploring space but are also forging a path for the future of humanity as a spacefaring civilization.

Chapter 7: Ancient Astronomy

Introduction

From time immemorial, humanity has been craning necks skywards, engaging in the profound act of stargazing, leading to the birth of astronomy that has had an indelible impact on cultures and religions since prehistoric times. These ancient observers of the cosmos utilized their celestial insights not just for philosophical musings but as practical tools to devise calendars that charted the passage of time and to navigate the vast oceans before the compass was conceived. Civilizations such as the Babylonians, Greeks, Chinese, and Mayans, excelled in the field, making astoundingly accurate predictions and detailed celestial charts. They left behind awe-inspiring relics like Stonehenge, the Antikythera Mechanism, and the grandiose pyramid of Chichen Itza that echo their astronomical prowess. Their fascination wasn't limited to the Sun and Moon alone; they mapped constellations, imbuing them with stories and legends that form the tapestry of ancient mythologies across the world.

Astronomy in Art and Literature

Throughout the ages, the ethereal beauty of the cosmos has been a boundless source of inspiration for artists and writers, evoking emotions and igniting creativity across civilizations. The iconic imagery of starry night skies has been immortalized by painters and photographers, each attempting to capture the enigmatic beauty of the heavens. Poetry, music, and narratives have borrowed celestial elements to craft metaphors that reflect upon life, existence, and the human experience. In these creations, the infinite reaches of space often provoke contemplation about our place in the universe, igniting both existential angst and wonder.

The field of science fiction, in particular, thrives on the advancements and conjectures of astronomy, envisaging future scenarios that ponder our species' fate and potential amidst the stars. Artists, utilizing modern theories and astronomical discoveries, have pioneered novel methods of visualizing data, blurring the lines between scientific endeavor and artistic expression.

Quotes About Astronomy's Influence

The allure of the cosmos has inspired a variety of voices from various walks of life, each providing unique insights into humanity's relationship with the stars. Whether you are a scientist, philosopher, artist, or laborer, you cannot help but be inspired by the mysteries of our universe. Here is a compilation of thoughtful reflections on the theme:

- "The cosmos is within us. We are made of star-stuff. We are a way for the universe to know itself." - Carl Sagan
- "We are the children of a stardust paradise." - Michio Kaku
- "The universe is not a place where we live, it is a process in which we are part of." - Neil deGrasse Tyson
- "Every atom in your body came from a star that exploded. And, the atoms in your left hand probably came from a different star than your right hand. It really is the most

poetic thing I know about physics... You are all stardust." - Lawrence M. Krauss
- "The Sun is a miasma of incandescent plasma." - They Might Be Giants, from the song "Why Does the Sun Shine?"
- "In learning about the universe, we learn about ourselves." - Neil deGrasse Tyson
- "Looking up into the night sky is looking into infinity— distance is incomprehensible and therefore meaningless." - Douglas Adams
- "The history of astronomy is a history of receding horizons." - Edwin Hubble
- "Space exploration is a force of nature unto itself that no other force in society can rival." - Neil Armstrong

These quotes encapsulate the profound interconnectedness between us and the universe, a reminder of our shared origins in the stars and our enduring quest to explore and understand the infinite expanse that surrounds us. They reflect a collective yearning to uncover the secrets of our cosmic journey and to find our place within the grand tapestry of existence.

Archeoastronomy

Archaeological findings continue to unravel our ancestors' deep connection to the stars. Remarkable finds such as Aboriginal rock paintings and petroglyphs reveal depictions of supernovae events, indicating these ancient artists were chroniclers of celestial happenings. The Nebra sky disk, a Bronze Age artefact unearthed in Germany, astounds modern observers with its detailed representation of the cosmos, solidifying its place as the earliest known realistic depiction of the night sky. Excavations at sites like Göbekli Tepe have led to the astonishing realization that our prehistoric forebearers constructed astronomical observatories millennia before Stonehenge ever stood. Archaeology has confirmed that numerous ancient monuments and megalithic structures were not random in their orientation but deliberately aligned with celestial events such as solar solstices, equinoxes, and lunar cycles, integrating their understanding of astronomy into their spiritual and everyday lives.

Modern Portrayals of Astronomy

In the modern era, the portrayal of astronomy through various media has mirrored historical achievements and captivated public imagination. Iconic events like the Apollo Moon landings of the 1960s have been enshrined in cultural memory, influencing a myriad of artistic and media depictions. Similarly, recent groundbreaking feats such as the first photographs of a black hole have fueled a new wave of artistic endeavor. Planetariums, with their immersive capabilities, offer digital vistas that simulate the grandeur of the universe, allowing individuals to voyage across the cosmos while seated. The field of space art is ever-evolving, with artists harnessing both traditional and digital mediums to interpret and represent the wonders of the universe, infusing pop culture with images that juxtapose the surreal and the familiar, inviting viewers to contemplate their place in the grand canvas of space. Science fiction writers, inspired by these advances, continue to push the envelope, incorporating concepts like warp drives, wormholes, terraforming, and Dyson spheres into their narratives. These stories fuel the collective yearning for a future where humanity transcends its planetary cradle to become a true spacefaring civilization.

Chapter 8: Upcoming Space Telescopes

Introduction

The future of astronomy is poised on the cusp of a new era of discovery with a fleet of advanced space telescopes awaiting deployment. The NASA James Webb Space Telescope, with its launch scheduled in 2021, will offer an unparalleled window into the cosmos via infrared observations. Its primary missions include probing the atmospheric components of distant exoplanets, capturing the light from the universe's first galaxies, and providing new insights into the narrative of cosmic evolution. The late 2020s will welcome the Nancy Grace Roman Space Telescope, destined to conduct extensive wide-field surveys that will delve into the enigmatic realms of dark matter and dark energy. The roster of future observational marvels further extends to SPHEREx, which will scrutinize galaxies for signatures of water and life-essential organic compounds, and LUVOIR, the proposed large optical observatory with the ambitious goal of directly imaging exoplanetary systems. These instruments represent a monumental leap forward in our capabilities, each engineered to uncover layers of cosmic phenomena that have thus far eluded our grasp.

Asteroid Detection and Prevention

As our awareness of celestial dangers grows, so does the importance of asteroid detection and prevention. Ongoing programs like Pan-STARRS and ATLAS are relentless in their pursuit to catalog near-Earth objects, especially asteroids larger than 140 meters, aiming to detect 90% of such bodies. In the face of a tangible impact threat, humanity's contingency plans include deploying kinetic impactors, designed to nudge threatening asteroids off a collision course with Earth, and as a last resort, the controversial yet potentially necessary option of nuclear disruption. The endeavor to refine these planetary defense measures is not merely academic but a crucial step towards the safeguarding of Earth's inhabitants.

Quotes About the Future of Astronomy

The field of astronomy stands on the precipice of a new era, with the vastness of the universe beckoning us towards untold discoveries and insights. As we peer deeper into the cosmos, the future of this science is shaped by the anticipation of what lies beyond our current horizon of knowledge. The following quotes encapsulate the spirit of what we can expect and dream about the future of astronomy:

- "Exploration is in our nature. We began as wanderers, and we are wanderers still." - Carl Sagan
- "Astronomy compels the soul to look upward, and leads us from this world to another." - Plato

- "The important thing is not to stop questioning. Curiosity has its own reason for existing." - Albert Einstein
- "Two possibilities exist: either we are alone in the Universe or we are not. Both are equally terrifying." - Arthur C. Clarke
- "For my part I know nothing with any certainty, but the sight of the stars makes me dream." - Vincent Van Gogh
- "It is not in the stars to hold our destiny but in ourselves." - William Shakespeare
- "The more clearly we can focus our attention on the wonders and realities of the universe about us, the less taste we shall have for destruction." - Rachel Carson
- "To confine our attention to terrestrial matters would be to limit the human spirit." - Stephen Hawking
- "The sky calls to us. If we do not destroy ourselves, we will one day venture to the stars." - Carl Sagan
- "Space is the ultimate frontier for the future of mankind." - Michio Kaku
- "We are all in the gutter, but some of us are looking at the stars." - Oscar Wilde

These reflections from thinkers, scientists, and philosophers offer a glimpse into the profound impact that astronomy will continue to have on humanity. They remind us that our exploration of the cosmos is not just about the acquisition of knowledge, but also about the profound implications for understanding our place in the universe and the future of our species.

Cutting-Edge Astronomy Research

Astrobiologists are meticulously studying extremophiles—organisms thriving in Earth's harshest conditions—to model the potential for life in extraterrestrial settings. Gravitational wave astronomy, a nascent yet revolutionary field, is forging new pathways in testing general relativity and understanding the expansion of the cosmos by detecting ripples in spacetime.

Researchers studying the cosmic microwave background are engaging in high-precision measurements to unravel the mysteries surrounding the universe's birth and its subsequent evolution. Additionally, the development of next-generation extremely large telescopes, epitomized by the Thirty Meter Telescope, promises to escalate our direct imaging capabilities, potentially revealing exoplanets in striking detail.

Long-Term Visions for the Field

Astronomy's long-term aspirations stretch from our planetary neighbors to the far reaches of the galaxy. The colonization of Mars is envisioned as a stepping stone towards the broader human settlement of the cosmos. Ambitious plans for interstellar probes are being conceptualized to reach neighboring stars like Proxima Centauri, with intentions to communicate their findings back to Earth. The field of propulsion is witnessing imaginative advances with the theoretical development of light sails, nuclear pulse engines, and antimatter drives, each with the potential to revolutionize deep space travel. Mega-scale engineering projects, once the realm of science fiction—such as Dyson spheres and stellar engines—begin to enter the realm of plausibility for technologically advanced civilizations. The culmination of these advancements could, in the far future, result in a cosmos teeming with interstellar economies, communities traversing star systems, and the proliferation of intelligent life spreading its influence across the galaxy.

That concludes our journey into the fascinating world of astronomy, and the secrets it has revealed about our universe.

Now, stay tuned for a FREE excerpt from one of our bestselling books, AI for Smart Pre-Teens and Teens Ages 10-19. I hope you will consider adding it to your library.

FREE SAMPLE from this book coming up NEXT!

Get on the leading edge of the AI Revolution!
For pre-teens, teens and older

- AI demystified: learn the fundamentals behind AI, including machine learning, neural networks, and deep learning
- Get inspired by industry leaders like Sam Altman, Elon Musk and Satya Nadella
- Help drive the responsible and beneficial use of AI
- Perfect travel companion or gift

Follow Dr. Leo Lexicon on Twitter/X
X @LeoLexicon

FREE SAMPLE Chapter (Ch.5 From the Mouths of Experts)

Artificial intelligence is shaping the future, but it didn't happen overnight. Over the years, the development of AI has benefited from the genius of many people. We will hear from some of the top innovators who helped shape AI in this chapter. Prominent individuals from a range of businesses have different viewpoints on the possible implications and difficulties of artificial intelligence (AI), a topic that is fast growing. While this list is not exhaustive, it is always important to identify and understand the viewpoints of key people in any domain. Here is a more detailed look at their opinions:

Sam Altman, OpenAI's CEO

Sam Altman is a well-known name in the AI field, and he has a very upbeat outlook on how AI could change the world. According to Altman, AI has the potential to outperform even the revolutionary effects of early breakthroughs like electricity and fire. He imagines a time when AI will be used as a potent weapon against serious global problems like sickness and poverty. In Altman's opinion, the proper use of AI may change industries, boost productivity, and enhance people's quality of life all across the world.

Altman does acknowledge that there are considerable hazards associated with such enormous potential. He is fully cognizant of the potential ethical, social, and economic difficulties that AI may present. Altman highlights the significance of ensuring that the advantages of AI are available to everyone, regardless of socioeconomic class or geographic location, as AI develops. In order to prevent escalating already existing inequities, he supports regulations that encourage a just and equitable sharing of AI's benefits.

Elon Musk, Founder of Tesla

Elon Musk has a long connection with AI, and he was in fact one of OpenAI's co-founders, but later dropped out of the company, long before it released ChatGPT. Elon's views suggest that he approaches AI with more caution. He acknowledges that AI has the potential to improve a variety of industries, but he is also gravely concerned about the risks that could arise from unrestrained AI research. He has even compared AI to nuclear weapons in an effort to warn that, if not properly governed, it could endanger humanity.

The lack of regulatory control in AI research is Musk's main worry. He thinks that in the absence of appropriate rules and regulations, AI systems might be used without sufficient safety precautions, which could have unforeseen repercussions. Musk is a supporter of strict governance and ethical frameworks that put safety and human values first in order to ensure that AI is developed ethically. In order to ensure that AI research and application are carried out with the highest caution and responsibility, he argues for the creation of institutions and policies.

Musk recently announced the creation of a new company called X AI devoted to responsibly advancing artificial intelligence. The company will focus on creating transparent AI systems that can articulate their decision-making processes in a way humans can understand. Musk said X AI will allow AI developers "to see how the AI thinks and why it makes certain decisions." He hopes X AI will set a new gold standard for ethical, thoughtful AI design that other companies will follow, leading to AI that truly augments human abilities. Given Musk's profile, X AI is sure to quickly become an influential player in the high-stakes world of AI safety research.

Yoshua Bengio, Deep Learning Pioneer

Deep learning pioneer Yoshua Bengio provides insight into the state of AI today and its limits. He calls AI "narrow and brittle," emphasizing how most AI systems do well at performing narrowly defined tasks but fall short when it comes to generalizing knowledge and responding to novel circumstances.

Artificial general intelligence (AGI), which refers to AI systems with the capacity to reason and comprehend the world at a level equivalent to human intelligence, is a goal strongly supported by Bengio. He thinks that in order to attain AGI, it is crucial to learn more about how the human brain works and how neural networks and biological processes contribute to intelligence.

Bengio believes that cognitive and neuroscience research is essential for directing the creation of AI systems that can learn, generalize, and adapt just like people. In order to create more adaptable and flexible AI systems, he promotes multidisciplinary collaborations between specialists in AI and other scientific domains.

"The Godfather of deep learning" Geoffrey Hinton

One of the pioneers of deep learning, Geoffrey Hinton is internationally recognized for his contributions to the development of AI. Hinton's viewpoint focuses on the necessity of updating AI datasets and algorithms to produce more intelligent machines. He berates modern AI models for not having a thorough comprehension of reality. Although AI has made great progress in a number of specific tasks, Hinton contends that these systems frequently lack the reasoning and knowledge generalization skills necessary for attaining real intelligence. He urges the use of fresh methods in AI research that could help us comprehend the fundamentals of human intelligence better.

Hinton imagines a time where AI helpers can relate to people as dependable friends and have empathy for them. He thinks that for AI to succeed, it must advance beyond simple pattern recognition and gain comprehension of context, emotions, and intentions. In Hinton's perspective, AI can be a real collaborator who can connect with people on a deeper level rather than just a tool for particular jobs.

Fei-Fei Li, Co-director of Stanford's AI Lab

Fei-Fei Li, a well-known researcher and lecturer in the field of AI, believes that AI has enormous potential to enhance human potential and improve life. She is aware of how AI can revolutionize many industries, such as healthcare, education, and environmental sustainability. AI is a useful tool for tackling complex problems and improving scientific study because of its capacity to process enormous volumes of data and spot patterns.

However, Li stresses that it is crucial that AI is developed in an ethical and responsible manner. She is an advocate for greater diversity in the AI community since it can lessen the negative biases present in the technology. To prevent sustaining societal disparities, it is essential to ensure inclusivity and justice in AI applications.

Li is a strong supporter of the viewpoint that AI should be created with an emphasis on enhancing rather than replacing human capabilities. She thinks AI systems ought to be created so they may coexist peacefully with people, boosting their skills and enabling them to make wiser judgments. AI may be used as a potent tool to address global concerns and enhance the quality of life for all people by keeping humans in the loop and prioritizing human-AI collaboration.

Andrew Ng, Google Brain Co-Founder

The well-known AI researcher and entrepreneur Andrew Ng is certain that AI has the power to fundamentally alter the healthcare industry. He views artificial intelligence as a useful tool that can help with earlier and more precise disease diagnosis, perhaps saving lives. AI can help doctors make better judgments and improve patient outcomes by analyzing massive volumes of medical data and looking for trends.

Ng warns against overestimating AI's potential, though. While AI has demonstrated tremendous accomplishments in certain jobs, it still lacks the human ability for common sense and general intelligence. Due to this constraint, AI might be excellent in specific fields yet struggle to comprehend complicated real-world situations that call for human-level comprehension and reasoning.

Demis Hassabis, DeepMind's CEO

Demis was the leader of the team that developed AlphaGO, the AI that famously defeated the GO world champion, Lee Sedol. Demis imagines a time when AI and people work in unison to solve the world's problems. He understands that AI is particularly good at digesting large volumes of data and making judgments based on that data. Humans, on the other hand, contribute special abilities like compassion, imagination, and intuition to the table. Hassabis thinks we can better tackle complicated issues if we combine the analytical strength of AI with human qualities.

Under Hassabis' direction, DeepMind has been aggressively examining how AI could help professionals in a range of industries, including healthcare and scientific research. Hassabis's idea of AI-human synergy is demonstrated through the company's partnerships with medical organizations to develop AI systems that assist in disease diagnosis and drug research.

Cynthia Breazeal, Creator of Jibo

A pioneer in the field of human-robot interaction, Cynthia Breazeal focuses on developing social robots that can comprehend and relate to people emotionally. She is adamant that in order for society to accept AI, it must be more than just a machine that performs tasks; it must exhibit human-like traits that enable sincere emotional interactions.

The goal of Breazeal is to create AI-enabled robots that can perceive and react to human emotions, allowing them to develop deep connections with their users. Such sympathetic AI can be used in a variety of industries, including healthcare, where social robots can help patients emotionally, and education, where they can effectively engage and motivate students.

Eliezer Yudkowsky, Machine Intelligence Research Institute

Eliezer Yudkowsky is a fervent supporter of AI ethics and safety. He issues a warning regarding the dangers that could arise from the creation of complex AI systems. A lack of appropriate prudence and ethical standards could have unforeseen repercussions or even threaten humanity's existence.

To ensure that AI systems behave ethically, Yudkowsky highlights the significance of matching AI's objectives with human ideals. Yudkowsky thinks we can prevent situations in which AI can unintentionally injure people or behave against their best interests by giving safety a higher priority in AI research and development.

Ilya Sutskever, Co-Founder of OpenAI

Ilya Sutskever is one of the co-founders of OpenAI. He asserts that artificial intelligence (AI) may overtake human intelligence within a few decades. One of humanity's biggest concerns may be how to transition to digital superintelligences.
Sutskever underlines the significance of incorporating human values into the design of new AI systems. As AI develops into superintelligence, it is essential to make sure that these potent agents act in accordance with human interests and prevent any unanticipated negative outcomes.

Satya Nadella, CEO of Microsoft

In order to avoid relying on "black box" solutions, Satya Nadella urges businesses to embrace AI as a core skill. Satya Nadella is adamant that AI will change every industry. In the future, he sees AI influencing many facets of corporate operations, from boosting productivity to tailoring customer experiences.

Nadella is dedicated to democratizing AI so that it is available to people and organizations of all sizes. He sees a world where AI platforms and tools are accessible to everyone and are easy to use, allowing more people to use AI to solve issues and spur creativity.

Yann LeCun, Facebook AI Research

Yann LeCun Leading expert in AI Yann LeCun is a proponent of AI systems that enhance and supplement human abilities. LeCun contends that creativity and common sense, which present AI systems lack, are essential components of intelligence.

LeCun advises concentrating on creating AI systems that can collaborate with people in a positive way in order to produce AI that can actually improve human intelligence. In order to provide AI systems the ability to learn more independently and comprehend complicated real-world settings better, research is being done in areas like reinforcement learning and unsupervised learning. LeCun sees a time when artificial intelligence (AI) bridges the gap between human and machine intelligence, enhancing human capabilities and assisting in the solution of humanity's most pressing problems.

Mo Gawdat, Former Chief Business Officer, Google X

Mo Gawdat, also the author of a popular book, Solve for Happy, has an optimistic yet cautious perspective on artificial intelligence. He believes AI has immense potential to automate jobs and tasks, freeing up human creativity and time for higher pursuits. However, Gawdat also recognizes the risks of superintelligent systems and the need to align advanced AI with human values and ethics. He advocates for policies to smooth the transition and distribute the benefits, along with designing AI thoughtfully with emotional intelligence and compassion in mind. Overall, Gawdat argues we can harness the power of AI to improve lives if we steward it carefully, create wise governance, and democratizing access to shape an ethical, benevolent AI that enhances human potential while mitigating the existential risks.

A Final Word

I truly appreciate your participation in this unique journey through the wonders of astronomy. Do stay tuned for a future release titled "Astronomy Nerd: Quizmaster Edition", that will feature challenging quizzes that test your knowledge, similar to the other books in the Quizmaster series.

If you liked this book, please help me spread the word by:

- Leaving a 5-star review on Amazon
- Telling your siblings, classmates, friends and relatives about this book
- Recommending this book to your teacher, coach or educator, and
- Sharing your thoughts on social media

I hope you also liked the free sample from the book, AI for Smart Pre-Teens and Teens Ages 10-19. Do check out our other exciting titles and stay tuned for new and exciting releases from Lexicon Labs. Some of them are highlighted in the pages that follow.

Don't forget to sign up to our newsletter to be informed of future titles and updates, and download your FREE POSTER. The signup link is https://mindzen.squarespace.com/lex

I wish you lots of good luck and new adventures!

Dr. Leo Lexicon

Thank you for reading this book. Please write a review, and share with your friends on social media if you enjoyed this title.
We are counting on you to spread the word!

If you liked this book, you will also enjoy new and upcoming titles from our Great Explorers series about extraordinary adventurers like Ferdinand Magellan, Ernest Shackleton, Roald Amundsen, Marco Polo, and others!

Explore more titles from Lexicon Labs in the pages that follow.

Don't forget to sign up to our newsletter and download your FREE poster print! Go to https://mindzen.squarespace.com/ and sign up today!

Explore the lives of great innovators, Scientists, Leaders, Artists and Explorers...Stay tuned for additional titles coming soon!

Learn the basics of Coding and program in Python.
No prior knowledge required!

Meet our bestselling titles on AI
BOOKS FOR CURIOUS MINDS

- Structured introduction to the building blocks of AI
- Review of major milestones in AI history
- Meet the leading inventors and their key innovations
- AI concepts explained in a simple, easy-to-understand format by a Bay Area educator
- Resources for puzzles, games, and coding
- Perfect travel companion or gift

Follow Dr. Leo Lexicon on Twitter/X

 @LeoLexicon

LEXICON LABS

Learn all about starting and growing a business <u>as a teenager</u>

Explore the Future of Quantum Computing

FUN BOOKS FOR TRIVIA NIGHT

COLORING BOOKS

TEST YOUR INNER NERD!

COLORING BOOKS

Explore the Future of Quantum Computing

Check out our fun, auto-themed coloring books

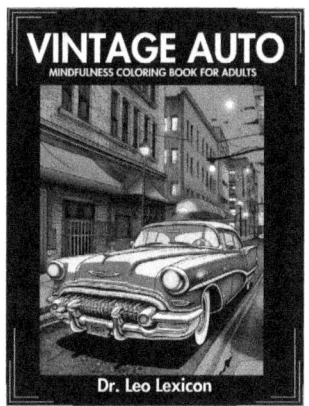

MASTER PYTHON!
The Swiss Army Knife of Programming

SCAN ME

- Start from the basics, advance to intermediate level
- Fun coding projects and examples
- Python concepts and usage explained in a simple, easy-to-understand format by a Bay Area educator
- Build new projects using your knowledge base
- Learn about career paths that need your Python skills

Follow Dr. Leo Lexicon on Twitter/X

X @LeoLexicon

LEXICON LABS

AI FOR YOUNGER READERS
Get on the leading edge of the AI revolution!

- Perfect for readers Ages 6-9
- Structured introduction to the building blocks of AI
- AI concepts explained in a simple, easy-to-understand format by a Bay Area educator
- Resources for puzzles, games, and coding
- Perfect travel companion or gift

Follow Dr. Leo Lexicon on Twitter/X

 𝕏 @LeoLexicon

Explore the Quizmaster Collection
Know your facts, dazzle your friends,
and discover your inner nerd!

Discover More Bestselling Titles
from Lexicon Labs!

SCAN ME

www.ingramcontent.com/pod-product-compliance
Lightning Source LLC
Chambersburg PA
CBHW070804290526
45795CB00002B/621